The Universe

Dr Greg Brown

Royal Observatory Greenwich
Illuminates

First published in 2022 by Royal Museums Greenwich, Park Row, Greenwich, London, SE10 9NF

ISBN: 978-1-906367-89-3

At the heart of the UNESCO World Heritage Site of Maritime Greenwich are the four world-class attractions of Royal Museums Greenwich – the National Maritime Museum, the Royal Observatory, the Queen's House and *Cutty Sark*.

rmg.co.uk

A CIP catalogue record for this book is available from the British Library.

Typesetting by ePub KNOWHOW
Cover design by Ocky Murray
Diagrams by Dave Saunders
Printed and bound by GRAFO

About the Author

Dr Greg Brown is an astronomer working at Royal Observatory Greenwich. In his time in research at the University of Warwick, he studied some of the largest explosions in the Universe and the supermassive black holes hiding in distant galaxies. Combining a love of science, comedy and acting, Greg moved into science communication, where he has been eliciting anguished groans from his audiences ever since.

Entrance to the Royal Observatory, Greenwich, about 1860.

About Royal Observatory Greenwich

The historic Royal Observatory has stood atop Greenwich Hill since 1675 and documents over 800 years of astronomical observation and timekeeping. It is truly the home of space and time, with the world-famous Greenwich Meridian Line, awe-inspiring astronomy and the Peter Harrison Planetarium. The Royal Observatory is the perfect place to explore the Universe with the help of our very own team of astronomers. Find out more about the site, book a planetarium show, or join one of our workshops or courses online at rmg.co.uk.

Contents

Introduction

The Universe.

As far as topics go, this is a pretty big one.

Astronomy has a tendency to deal with some of the largest objects there are, but when the subject of discussion is the entirety of known existence, not to mention its history and future to boot, we are quite literally talking on an entirely different scale.

Thankfully, there is a whole field of astronomy dedicated to understanding the processes that have shaped our Universe. It's called cosmology – the study of the origin, evolution and structure of the Universe, from its very beginnings through to its eventual demise – though let's not get ahead of ourselves.

People have wondered about the nature of existence since it was possible to wonder at all. Many of the theories and beliefs that have popped up over the centuries have long been left behind, yet it is surprising how young our understanding really is. A handful of people alive today may still remember when the prevailing theory was that the Milky Way was the only galaxy in the Universe and space was a lot smaller than we now believe it to be.

Time, and a great deal of effort, has shown us that the Universe is almost unimaginably large, incredibly old, and simultaneously practically an infant. The theories of physics that we are used to in our daily lives are specific cases that fall apart at times when looking at the very big and the very small – both of which are extremely important when discussing the Universe as a whole. The result is a Universe that defies simple explanation and whose laws seem to also defy logic. In short, it's weird.

In these pages, we'll talk about the humble origins of the Universe, from the Big Bang to the first stars, before moving onwards to consider it as it is today. We'll discuss how the imprints of the earliest moments in the Universe are still visible, dictating its grand structure for the rest of time. And we'll look ahead to see how the Universe will evolve from now to the fascinating, if macabre, subject of its end. But more important than all of that, we'll talk about how scientists, astronomers and cosmologists are piecing everything together with scraps of information hidden in the most elusive parts of the Universe, parts that no human eye could ever see. A glossary of useful terms, shown as bold in the text, can be found at the end of the book.

I've got more than $10^{10^{120}}$ (that's a one with one novemtrigintillion zeros after it) years of history to cover in a pocket guide...

I'd best get started.

Our Ever-Expanding Map of the Universe

In many ways, the story of early astronomy, and a fair bit of not so early astronomy, was the story of finding our place in the Universe. We were drawing a map of the entirety of existence and placing ourselves on it... somewhere.

As anyone but the boldest (or most foolish) of us knows, the way to start a jigsaw puzzle is to put together the edge pieces first. Similarly, when making a map, it's a good idea to know how big it needs to be before you start drawing. So, the question for early astronomers became 'how big is the cosmos?' Where do we place the edge of the Universe?

We begin, as most histories of science do, with the Ancient Greeks and with one in particular: Aristotle. Born in 384 BC, he is widely regarded as one of the greatest thinkers of his age. An eminent philosopher, even a scientist of sorts over 2,000 years before the term would be coined, his writings covered almost every conceivable topic under the sun and beyond, from ethics to economics, politics to poetry. He was taught by the great Plato himself and became tutor to Alexander the Great when he was still Alexander the Mostly OK. His work was so respected and revered that for almost two millennia much of our understanding of the Universe was built on its foundations.

This was a bit of a problem, however, as a lot of what he came up with was, well, wrong. Unfortunately, Aristotle's fame, which approached legendary status in the Middle Ages, made him a difficult person to argue against – that and the fact he was dead.

His views of the structure of the Universe caused countless issues. Like many before him, he viewed the Earth as a fixed place within a Universe that revolved around it. The Moon, Sun and planets orbited the Earth, as did the distant stars. But even those were not so far away, just a little beyond the orbit of Saturn, the furthest known planet at the time (Figure 1). Outside of that, perhaps nothing, a void not worth concerning yourself with.

Aristotle's model of the Universe was a small one by modern standards – a fraction the size of our Solar System as we know it today. What's more, this map was far from complete by any cartographer's standards. With apparently no way to determine the distances to things, a true scale could not be placed on it.

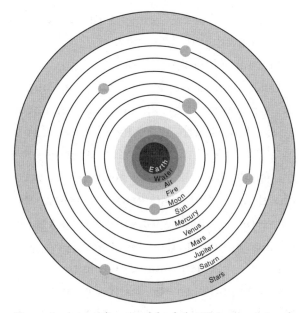

Figure 1: Aristotelian model of the Universe. Aristotle had specific places for each of his four earthly elements and a cosmos made from the fifth. His map has some features we would recognise today, like the general order of the planets, and the Moon being closer to the Earth than the Sun, but otherwise there were many issues that wouldn't be identified or understood for more than 2,000 years.

At least, not until Aristarchus, another ancient Greek, this time a mathematician, came along. Working a little after the death of Aristotle, he attempted to make some of the first real measurements of space and the bodies that it contained.

He started with the Moon.

Once every couple of years, our night skies are graced by an unusual alignment of the Sun, Earth and Moon, known as a total lunar **eclipse**. The Moon passes into the shadow of the Earth and turns a deep blood red. But just before totality, the curve of the shadow of the Earth is visible on the Moon's surface. Aristarchus realised that he could determine how big the Moon was compared to the Earth by assuming this shadow was the same size as the Earth itself. Then, by measuring how big the Moon appears in our skies and using the simple fact that objects appear smaller the further away they are, he found the Moon was roughly 25 Earth widths distant – not bad considering the true result is about 30 (Figure 2).

The far more difficult task of determining the distance from the Earth

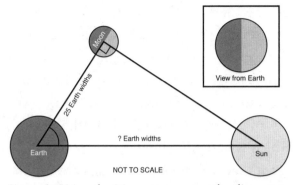

Figure 2: Using the Moon to measure the distance to the Sun. When Aristarchus was looking to measure the Solar System, he realised he could do it using the Moon. When the Moon is in its exact quarter phase, that is when half of it is lit up from our point of view, the angle from the Earth to the Sun at the Moon's surface must be 90 degrees. Knowing the distance to the Moon already meant the whole triangle could be solved by measuring the angle from the Moon to the Sun from the Earth's surface. His result, though some way from the true value, nonetheless started to put a scale on the Solar System.

to the Sun required identifying the exact time of the quarter moon, the phase of the Moon where half of its surface is lit up from our point of view, and then measuring the angle from the Moon to the Sun. At quarter moon, the angle of the Sun's light on the Moon must be exactly 90 degrees, so the resulting right-angled triangle could be solved completely using trigonometry. Aristarchus' measurements implied that the Sun was 20 times more distant than the Moon, which is actually about 20 times too small given that the modern value sits around the 400 mark. Nonetheless, he had shown that the Solar System was pretty vast, at least several hundred times the size of the Earth, and already humanity was smaller than it might like to believe.

Because he knew the Sun was much more distant than the Moon and yet appeared to be the same size, Aristarchus realised the Sun must be vast and briefly considered the possibility that the Sun might be the centre of the Solar System, not the Earth. To say that this idea didn't take for a while would be one of the greatest understatements in the history of cosmology...

As useful as this was, it still didn't place a true scale on the known Universe. Aristarchus may have established that the Moon was about one-quarter the width of the Earth, but what that meant in metres or feet was impossible to say – partly because those units of measurement didn't exist yet, but mostly because no one knew how wide the Earth was.

The one to solve this would be Eratosthenes, yet another famous Greek mathematician. He noticed that at noon on the summer solstice the Sun's light would fall straight down a deep well in Syene, now Aswan, in Egypt. But at the same time in Alexandria, some 800 kilometres (500 miles) away, the Sun was lower in the sky. By assuming the difference was entirely thanks to the curvature of the Earth, he calculated that the circumference of the Earth was about 40,000 kilometres (25,000 miles). This was an incredible result, within

1 per cent of the modern value, but it also gave astronomers an extremely important tool: a way to measure the rest of the Solar System, and perhaps even beyond.

Have you ever wondered why humans have two eyes?

This question will make sense in a moment, trust me.

While you may think it's to provide a backup in case something happens to one of your eyes, or perhaps simply for aesthetics, it's actually because having more than one perspective can be very beneficial. A close-by object will appear to jump against a more distant background when comparing one eye's view with the other. The closer the object, the bigger the shift. Your brain then uses a combination of this shift and yet more trigonometry to judge the object's distance, a process known as **parallax**. This has saved many a mug of tea from being swept onto the floor, though, of course, nobody is perfect.

The greater the distance between the two points of view, known as the baseline, the larger the distances you can triangulate. Similar methods have been used to map entire countries or measure the heights of mountains.

Now that the size of the Earth was known, two observers making simultaneous observations from vastly different locations might be able to push parallax all the way out to space, thanks to the massively improved baseline that would provide. That said, this would require a level of coordination that would be difficult to achieve until much later, not to mention accurate observations with telescopes, which wouldn't be invented until the early 1600s. Until then, measuring distances in space was beyond the ability of astronomers and the map of our Universe would remain frustratingly incomplete.

After almost 2,000 years of scant progress, the 16th and 17th centuries saw a rollercoaster of changes. Nicolaus Copernicus would devise his **heliocentric model** of the Solar System, with the Sun rather than the Earth at the centre and, less than 100 years later, around 1610, Galileo Galilei's early telescope observations would all but confirm this theory. Johannes Kepler would build upon this work, realising that the time the different planets take to orbit the Sun is dependent on their distance from it. Because he already knew how long those orbits were, all he needed was one single measurement of the distance from one planet to another to calculate the scale of the entire known Solar System.

This would be provided by Giovanni Cassini and Jean Richer working from the Paris Observatory and French Guiana respectively. They observed the planet Mars in 1672, measured its parallax and, with remarkable accuracy, could finally say how big the Solar System really was.

The distance from the Sun to Saturn, still the furthest planet discovered by that time, was incredible: some one and a half billion kilometres. But more impressive still was the implication for the stars. Try as they might, astronomers could not register even the smallest shift from either side of the Earth. The stars must have been so distant that their parallax was undetectable – the map of the Universe must be much bigger than anyone had previously thought.

The first successful measurements wouldn't come until the 1800s when Friedrich Georg Wilhelm von Struve, Friedrich Bessel and Thomas Henderson – Russian, German and Scottish astronomers respectively – each working independently, finally saw the almost imperceptible parallax wobble of the stars, not by observing from either side of the Earth as their predecessors had, but from either side of its orbit. With a baseline of some 300 million kilometres to work with, about 30,000 times larger than the Earth's width alone, they could push parallax just far enough to measure the distances to stars. The measurements still had to be incredibly precise, and each had to be made around six months apart to get both perspectives, but it worked.

Their results showed that even the few stars that were close enough to appear to wobble a bit as the Earth made its yearly voyage around the Sun were incredibly distant. The closest stars to Earth, the triplet we now refer to as Alpha Centauri A, B and Proxima Centauri, were found to be around 40 trillion kilometres (24 trillion miles) from the Earth. In modern-day astronomy we use a more palatable distance measurement called the light year, the distance that light, the fastest thing in the Universe, can travel in one Earth year. Given that this vast unit is around 10 trillion kilometres, this places the Centauri triplet about 4 light years away. Distances much beyond that would be difficult to determine until the advent of astrophotography around the turn of the 20th century, and even then, vast swathes of the stars would be too far away for even the huge baseline of the Earth's orbit to make them move noticeably in the sky.

Nonetheless, our map of space had grown even further. We had come far from Aristotle's somewhat cosy picture of the Universe, but there was still a long way to go.

In time, astronomers would develop further ways to measure distances in space, ones that had far greater ranges. Henrietta Swan Leavitt provided the next major rung on the ladder. She was one of a number of people, often women, known as 'human computers', who worked on the complicated and usually repetitive mathematical tasks that would one day be done by modern electronic computers. Part of her work around 1910 was to examine photographic plates of the night sky and catalogue each photographed star by brightness, relative position and more. In the process, she discovered that Cepheid variables, stars that regularly grow brighter then fainter as if they are attached to a faulty dimmer switch, could be used to measure the size of our entire galaxy, the Milky Way.

On a clear night well away from city lights, you can see the structure of our galaxy as a faint swathe of light across the sky. It's the combined glow of all of those stars, so distant that we can't see them as individual points of light without the aid of a telescope. With Leavitt's Cepheid variables and some carefully calculated calibrations, astronomers of the time came up with scales for this vast complex of stars numbering in the tens of thousands of light years. Today, we know it to be around 100,000–200,000 light years across, a vast structure of which the Solar System is just one tiny corner.

In fact, it was such a large structure that at the time many astronomers believed it must be the sum total of the Universe and that our galaxy was all there was! Others, however, disagreed, pointing at a group of objects known at the time as 'spiral nebulae'.

Nebulae, literally meaning 'clouds', are just that: cloudy regions in deep space. They come in lots of shapes, sizes and types, from those that hide the formation of new stars to those that are made of the remains of dead ones. These nebulae were assumed to be small parts of our galaxy, but spiral nebulae seemed different. One in particular, the Andromeda nebula, appeared to have more of a type of explosion, called a **nova,** than the entirety of the rest of the Milky Way combined! How could that be possible, unless the Andromeda nebula was in fact the

Andromeda Galaxy, an entirely different collection of stars separate from our own, but of a similar size to our Milky Way? It would mean that these 'island universes', as they came to be known, were extremely numerous and that our Universe might be filled with them. This would make our Universe far, far larger than we had previously thought.

Disagreement over the scale of the Universe grew so great that a live argument, held in the presence of some of the greatest scientific minds of the time, took place to try and settle it. It was called 'The Great Debate', because, when this is as exciting as academic discourse gets, you make a big deal out of it.

In reality, the argument settled nothing except who was the better debater, but it wouldn't be long before the correct answer emerged. Leavitt's Cepheid variables came to the rescue again as Edwin Hubble, whose name is now given to one of the most famous telescopes ever built, used them to show that those spiral nebulae were indeed at extreme distances, far outside the Milky Way. At millions of light years away, even for the closest of them, the Universe was now incredibly vast.

In just over 2,000 years the Universe had gone from being a sizeable, but modest enough region of creation, no larger than it needed to be and centred on our own Earth with the heavens spiralling around it, to a gaping near-infinite void, peppered with collections of billions of stars called galaxies, of which our own Milky Way was just one. The Earth was now an infinitesimally small speck within a much, much grander whole. And our map had even more holes in it than ever before.

We still didn't know where the edge was though.

Our Actually
Expanding Universe

Here's a question: why is the night sky dark?

At first glance, there may seem to be an obvious answer: the main sources of light, the stars, are very distant and thus faint. But actually there's a very real reason to consider this question both complex and important.

Imagine an infinite Universe randomly filled with stars that has always existed and doesn't really change over time – stars may come, stars may go, but overall things stay the same. Because no direction is different from any other no matter where you look, eventually your line of sight will be broken by the surface of a star. In truth, you might have to look an incredibly huge distance if the stars are far enough apart, but it doesn't matter. Every point in the sky will be the surface of a star. The sky should be brightly illuminated in all directions – blindingly bright.

But it isn't.

One of many to realise the implications of this was Heinrich Olbers, a German astronomer working in the late 1700s and early 1800s. This observation has now become known as Olbers' Paradox. One way to solve a paradox is to realise that the situation that you have set up, the thought experiment, is somehow flawed. One or more of the assumptions that went into it must be wrong in order to make its conclusions appear nonsensical.

The most obvious assumption to do away with is the first one we state – an infinite Universe. If the Universe were small enough, the vast distances between stars would mean that not every line of sight ended in the blinding plasma of a stellar surface. But after the findings of Edwin Hubble and our subsequent understanding of the incredibly vast size of the Universe, as far as astronomers of the late 1920s and beyond were concerned, the Universe was practically infinite. Perhaps it was, perhaps it wasn't, but the Universe was certainly large enough that there was doubt that its size would solve the problem in and of itself. So, what could the solution be?

A few years after the Great Debate, Edwin Hubble, continuing to observe these distant galaxies in our skies, discovered something incredible using an effect we experience in our everyday lives without even realising it.

Astronomers were already well aware of something called the **Doppler effect**. If the source of a sound, like the roar of a car engine, moves towards an observer, the sound changes to a higher pitch because the sound waves are effectively being crushed together. If the source is moving away, the opposite happens – the waves get stretched out and the sound pitches down (Figure 3).

This happens with any type of wave – sound, light, Mexican – they all work. However, with light, instead of pitch you have colour, so moving a light source towards or away from an observer will cause the object to change its colour – a **redshift** if the object is moving away from you and a **blueshift** if it is moving towards you.

For this change in colour to be noticeable to the human eye would require incredibly high speeds that are rarely observed, so this effect is measured by more subtle means. Astronomers can look for emission or absorption lines, bright or dull spots in the spectrum of light

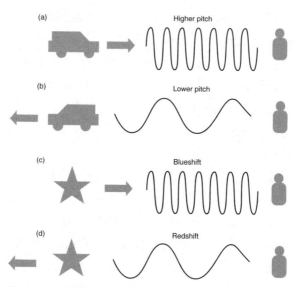

Figure 3: Thanks to the Doppler effect, the sound coming from a car engine is compressed when the car is approaching an observer (a) and stretched when moving away (b). This effect also applies to any source of light, such as a star or galaxy, with an approach causing a shift to bluer light (c), while receding causes a shift towards the red (d). Noticing that almost every galaxy in the Universe was redshifted and thus moving away from us was the big clue to understanding that our Universe was expanding.

produced by the various atoms in whatever is emitting or absorbing the light. Specific chemical elements produce recognisable patterns of lines that astronomers can identify. Because we know exactly what colour those lines should be, if we observe an object in space and see that line in the wrong colour, then we know it must have been shifted for some reason, presumably thanks to the Doppler effect. If you know how far along the spectrum the line has moved and whether it has shifted more towards the blue or the red side, then you can work out how fast the object is moving relative to you and in which direction.

When Hubble studied those distant galaxies, he found that, with only a handful of exceptions, they were all redshifted. Almost every galaxy in the Universe is moving away from us. What's more, he realised that the further away a galaxy was, the faster it was moving away too. A short time before, another astronomer, Georges Lemaître, had theorised the same possibility and so the rough relation they each came up with has now become known as the **Hubble–Lemaître Law**.

The inescapable conclusion seemed to be that almost everything in the Universe was moving away from us. Our Universe wasn't the static, unchanging realm we once thought it was, which had huge implications for Olbers' Paradox, not to mention the very nature of the Universe itself.

There is, however, one thing that may seem to be implied by this but is very much not. Despite apparent evidence to the contrary, we are still not the centre of the Universe.

It's easy to see why people might assume that. If everything in the Universe is streaking away from us at vast speeds, then whatever caused, or is causing, that motion must have happened, or be happening, here, right? Well... no.

There's a branch of physics known as **relativity**. It's the study of how different events appear from different perspectives. An event in physics is just something that occurs at a place and time – a concert could be an event, but so could someone switching on a lightbulb or a star exploding in deep space. Our position and motion can cause our perspective on these events to change and can make things look somewhat peculiar.

Imagine three cars on a smooth, straight road (Figure 4). The one in front is travelling at 60 mph, the one in the middle at 40 mph and the one bringing up the rear only 20 mph. Imagine that the road is smooth enough that the passengers in each can't actually feel themselves moving. What would the passengers in each car think was happening based on what they can see of the other cars?

The occupant of the rear one would see two cars moving away from them, one at 20 mph (the middle car) and one at 40 mph (the car in front). Because they can't feel themselves moving, they might assume that they are staying still and the others are the

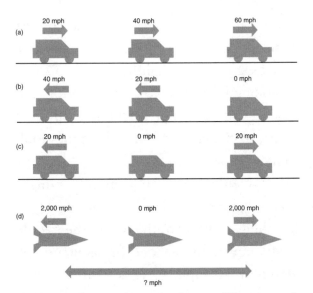

Figure 4: Three cars are moving at different speeds down the same road (a). Thanks to relativity, from the point of view of the front car the other two appear to be reversing away at high speed while it stays still (b), and from the point of view of the middle car both other cars appear to be rushing off in opposite directions while the middle car stays still (c). The occupants of each car think themselves in some 'special' place on the road. Complete the same experiment in space and the truth becomes clearer. Relativity makes it look as though the other spaceships are moving away while the middle one stays still, but with no fixed point from which to take a measurement, none of them really knows how fast the others are moving (d).

ones moving away. But, interestingly, the driver of the car at the front would also see two cars receding from view, one at a rate of 20 mph (the middle one) and one at 40 mph (the rear one). Despite the fact they are actually in the fastest car, they might start wondering what they haven't noticed about their car that would cause the others to be reversing away desperately at high speed.

In fact, the occupant of the middle car would also think that their vehicle is at rest with the other cars moving away from them. Each passenger would believe they were the epicentre of whatever was going on, when in truth none of them are. Similarly, while in our galaxy it seems that all other parts of the Universe are streaming away from us, making us look like some important focal point for all creation, aliens on another world in any of those other galaxies would be seeing our galaxy and, likewise, every other galaxy streaming away from them and struggling with their very own superiority complex.

There need not be a centre to the Universe at all in the usual sense. Remove the road and replace the cars with spaceships travelling in empty space and suddenly there is no way to determine which spaceships are moving, which are at rest and how fast they are all travelling except by comparing one to the next. In reality, this was true of the cars too. We may like to think of the road as static, but, on a spinning and orbiting planet in a mobile solar system and galaxy, who knows how fast the tarmac is actually moving? In the Universe, there is no fixed point against which to measure anything. Everything could be moving at incredible speed and we would have no way of knowing it. All those galaxies in the Universe appear to be moving away from us but we have no way of telling where the centre of that movement is or if there even is one.

Take any random large chunk of the Universe and compare it with another and they will look the same. This is known as homogeneity. Similarly, look in any random direction in the Universe and compare it with any other direction and they'll look the same too. This is known as isotropy. Together, the homogeneity and isotropy of the Universe are known as the **cosmological principle** and effectively it says nowhere in the Universe is 'special'. Humbling, perhaps, but it makes the maths so much easier. The galaxies aren't fleeing from any place in particular – they are just fleeing.

So how do all these fleeing galaxies solve Olbers' paradox? Well actually they do it twice!

Olbers' paradox assumes an infinite, static and everlasting Universe. While the first might still be true, the others no longer are thanks to this discovery.

We'll start with the Universe not being static. The Hubble–Lemaître law stated that the further a galaxy was from us, the faster it was retreating, and this could be seen through the redshift of the galaxy. What would happen if a galaxy was so far away that its light was redshifted out of our visible range and into the invisible infrared or radio light? Was that even possible?

In order for a galaxy's light to be pushed well out of our visible range it would have to be receding at more than the speed of light. But as we know from the work of Albert Einstein and his predecessors, not to mention every experiment that has followed, nothing in the Universe can exceed the speed of light. It is the absolute cosmic speed limit and, as such, is unbreakable, right?

Well, yes... but also no.

As far as our modern understanding of physics tells us, no object can move through space faster than the speed of light. There is a possible grey area about objects that have always moved faster than the speed of light but that's beyond the scope of this book. For everything else, if you start slower than light, you can never break that barrier while travelling through space.

But what if you aren't travelling through space? What if space is carrying you? Or more accurately, just appears to be carrying you?

Two people are standing a fair distance apart (Figure 5). Due to a longstanding disagreement that neither one can adequately explain, they hate each other and would frankly like to be much further apart. The simplest solution would be for one or the other to move away – but that would mean one would have to make concessions for the other, which neither can bear. The next simplest solution would be for both of them to turn their backs on each other and walk in opposite directions, but they would rather glare at each other

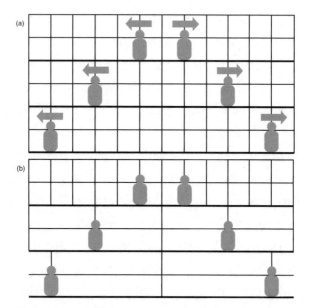

Figure 5: Expanding space. In scenario (a), two people decide to move apart. The distance between them grows because they are each moving in opposite directions. They are travelling through space. In scenario (b) the two people are fixed in place. Instead, the space between them itself expands; the distance between them grows as a consequence. They are not travelling through space – space is carrying them – and because of this they can appear to be moving at phenomenal speeds, far faster than the speed of light.

than turn away and, being safety conscious, they'd rather not walk backwards either.

A third solution is to grow space.

Granted this may seem a bit of an overengineered solution to a problem best solved with a combination of therapy and a bucket of cold water thrown over each of them, but it does fulfil all of their requirements. By growing the space between them, neither one is actually moving. This isn't semantics, they literally aren't changing their position – they are rooted in place – it's just that the space between them is getting bigger, so it looks like they are moving apart.

It turns out that this is precisely what is happening to the Universe. Those distant galaxies aren't moving away – it's just that all of space is getting bigger and that makes it seem as though space is dragging them away from us.

That's not to say that galaxies aren't moving at all – they are, sometimes at impressive speeds. When Hubble measured the speeds of galaxies and noticed that almost all were moving away, there were a few exceptions. The Andromeda Galaxy, for example, is actually moving towards us. This is because, being relatively close to the Milky Way, the apparent motion imparted to it by the expansion of the Universe is tiny compared to the true motion of the galaxy itself. Beyond our local neighbourhood, however, it's the expansion that dominates, carrying every other galaxy away from us. Even if the galaxy's real motion is towards us, it will be like a paddle boat struggling against the outgoing tide – ultimately it's going to recede whether its occupant wants to or not.

If this still does not make sense to you, may I be the first to congratulate on your perfectly functioning human brain. Whether we like it or not, humans have evolved from creatures whose entire existence has been within three-dimensional space, plus a bit of understanding of time. We have not experienced, in any meaningful fashion, space expanding or contracting and so our minds have never evolved to be able to envision it – imagining it happening is effectively beyond us. Even the most gifted astronomers and physicists have to put their natural instincts on hold when talking about complicated cosmological topics like the expansion of space. They can see the evidence of it, even if their moderately evolved ape brains won't accept it on a fundamental level.

So, what does this mean? Well, first it means that the redshifts Hubble had measured weren't really the Doppler effect at all. That only applies to objects that genuinely move. Go fast enough and the Doppler effect becomes squashed by **special relativity** – the thing Einstein came up with. The Doppler effect occurs at the moment of emission of the light (if it is the object that is moving) or at the moment of its detection (if it is the observer that is moving). But where galaxies are concerned, neither the observer nor the emitter is moving (much). Instead, the light itself is being stretched by the expansion of space while in transit. To distinguish this from redshifts and blueshifts due to actual motion, this effect is called **cosmological redshift**. It may look like a Doppler shift at low speeds, but it quickly becomes obvious that something else is happening at high ones.

The next implication is that, yes, galaxies can appear to exceed the speed of light and thus their light can be shifted out of our visual range. Generally speaking, this would mean that the more distant a galaxy was, the more its light would be shifted out of view and the dimmer it would appear. Go far enough and it would be effectively invisible. And so, the night sky is dark.

But there's more. The second way the expansion of space solves Olbers' paradox is a big one. The Universe isn't everlasting – it had a beginning.

Why does an expanding universe suggest it had a beginning? Well, if everything is currently rushing further and further away from us, what happens if you press rewind on the video? Everything starts rushing towards us, every point in the Universe bears down on every other, until everything is practically on top of itself. If taken to its extreme, this would imply a **singularity**, where all matter is crushed into an infinitesimal point with huge mass and zero size.

Now whether the Universe was actually a true singularity at any point is up for debate, but it was certainly smaller than it currently is – a lot smaller. Estimates vary but at one point it may have been that the entire dimensions of the observable Universe were smaller than a regulation football – soccer, not American…

The reason I said 'the observable Universe' rather than just 'the Universe' in that sentence is extremely important and, in fact, it's why the Universe having a beginning helps solve Olbers' paradox.

Light takes time to travel. It may be exceptionally fast, but it still takes 1.5 seconds to reach the Moon, 8 minutes to reach the Sun, 4 years to reach the next nearest star and 100,000 years to leave our galaxy. A universe that has always been around has had an infinite amount of time for light from incredibly distant stars and galaxies to reach us, but a universe that has a birthday, like ours, has only had its age. That means there is a limit to how far away we can see. At absolute most we can see how far light has travelled in the age of the Universe.

The Hubble–Lemaître law can give us an estimate on the length of that period. In an imaginary universe, in which the galaxies are moving away from one another twice as fast as they are in our own, the moment at the very beginning when all the galaxies were on top of themselves would have been half as long ago. Similarly, if the galaxies were moving away more slowly, then the beginning would have been earlier. Thus, a measure of the overall speed of these galaxies, known as the **Hubble constant**, can be used to find the age of the Universe. Today, that estimate stands at 13.77 billion years.

Naively you might think that means we can see 13.77 billion light years away but really it's just over three times that, again because the Universe is expanding.

Imagine a sprinter running the 100 metres. What if, just after the race began, someone started stretching the entirety of the track as though it were made of rubber? The distance the sprinter still has to run is being stretched, meaning they run a fair bit more than the 100 metres they intended to, but crucially the distance they have already run is also being stretched, so when they finish and look back it seems they've run even further still!

In exactly the same way, and because space has been expanding ever since the light started its journey, although a single photon of light has only physically travelled 13.77 billion light years, it will seem to have covered a distance of around 45 billion light years. Anything beyond that though is too far away. The Universe is too young for the light from those objects to have reached us yet. The observable Universe, the bit that we can see, is limited.[1]

And so, the night sky is dark.

But we do finally have an answer to the question we posed at the beginning of the last chapter. The edge of the map of our

---

[1] In fact, we can't see quite as far back as that either. There's an impenetrable (at least for light) barrier between us and the 'true' edge of the observable Universe – but more on that later…

Universe has been found – not, as it turns out, because that's where the Universe actually stops, but because we simply can't see any further.

What is there beyond our observable edge? We have no specific reason to believe the Universe outside our own observable bit of it is any different, at least not for a while. The edges of things tend to look different from the middle. For example, as you approach the coast, the sandy beach, salt spray and incessant noise of thieving gulls indicates you are coming close to the sea. But space looks pretty much the same in all directions and distances, exempting those differences that come from looking back in time to when the Universe was an infant.

Perhaps the space beyond our small corner is much vaster still, filled with more of the same as we see around us. Perhaps space is truly infinite, stretching on for ever. Or perhaps it is finite, yet unbounded, a phrase that requires a little explanation. It describes a Universe that only has a certain size but there's still no edge to it. Instead, travelling far enough in one direction in a straight line will end with you back where you began. It's like travelling off the right-hand side of the screen in a computer game and reappearing on the left-hand side – the world is only as large as the screen, but as far as the in-game character is concerned it doesn't have an edge. As weird an idea as this may be, some astronomers believe it to be a possible solution to some of the underlying weirdness in the Universe.

Whatever is the truth, we may well never know what is out there.

Hic sunt dracones – here be dragons.

While the phrase is something of an anachronism, it could work here in the same way as it has done on some maps of our world, when the picture of what was beyond remained unclear.

For now, though, because 'the observable Universe' is really all we can talk about with any certainty, it might as well be all of it to us – hence why I'll be referring to it simply as 'the Universe' from this point onwards.

The most incredible implication of discovering the expansion of the Universe, though, was the realisation that the Universe must have a beginning. Somehow the Universe went from being utterly tiny to incredibly vast and a mere 14 billion years was enough to create the cosmos we see today.

Actually, a lot of how our Universe is today was decided extremely early on, all the way back in...

The Very First Second

The Universe began.

If that's a little difficult to get your head around, you aren't alone. Granted, from ancient Egyptian mythology through to modern Judeo-Christian doctrine, there have been creation stories for as long as history can record. But to actually have scientific evidence that this statement is true is another thing entirely.

For the explanation to be that all of the stars, planets and other material, not to mention you and your archnemesis, all occupied basically the same space is frankly incredible. And, if you are anything like me, a little claustrophobia-inducing.

Of course, all that stuff wasn't the same as it is now – it was hotter for one thing; denser too, though perhaps that goes without saying. And none of it was in the slightest bit recognisable as what we see today.

Not until some time after the Big Bang.

You've probably heard the name before. You've probably even heard astronomers, physicists and the like refer to it as the beginning of our Universe. But science is rarely simple and the alleged beginning of the entirety of existence is, perhaps unsurprisingly, no exception.

The Universe may have had a beginning. It may even have been at the point we call the Big Bang. But, then again, it may not. We just don't know.

It would be far more accurate to say that the Big Bang was the beginning of the Universe as we know it today. Before that point, our understanding of physics falls apart and we can say no more. It may be that we are incapable of determining what came before, though that hasn't stopped people from trying. Theorists have concocted a veritable cornucopia of models – everything from colliding N-dimensional branes to pogoing cyclical expansions and beyond – but to understand them requires thousands of hours of study, an IQ in the region of 140 and a fondness for questions that have answers that we might never be able to confirm. That said, 'the beginning of the Universe as we know it today' is a bit unwieldy, so don't be surprised when I use 'the beginning' as shorthand further down the line.

So what was the Big Bang?

The term was coined in 1949 but was not commonplace until the 1970s and, as with many names in astronomy, it has led to some misunderstandings – the Big Bang was arguably neither big, nor a bang. This may be partly due to the fact that the name came from astronomer Fred

Hoyle, a supporter of an opposing theory. He favoured the 'Steady State theory' (later shown to be false), which suggested that new material was being created in the Universe as it grew. It seems, though, that his coining of the term 'Big Bang' was an earnest attempt to concisely describe it rather than a disparaging strawman.

The Big Bang is also often poorly described, though for completely understandable reasons. The Big Bang was not an explosion in any conventional sense of the word. It made no sound that we know of, required no explosive (chemical or otherwise) and did not have a central point from which material was ejected in all directions. Most depictions of it in film have space at a tiny point, followed by a burst of light and a rushing of stuff to fill an empty void. But, as I've already pointed out, we have no idea what is outside of our observable Universe, if anything, and in any case it is unhelpful to think of the expansion of space as something that displaced some void beyond our knowledge. Our Universe may have grown bigger, but it did so on the inside, like the interior of the TARDIS from Doctor Who gaining some extra rooms, not like a balloon expanding by robbing space from the world around it.

But, given that understanding that expansion is complicated, the analogy of an explosion is a fair one to get the basics across – hence why it is used so often.

What triggered this expansion? We don't know. What powered it? We don't know. How much of space beyond our Universe was also affected? We don't know. All we do know is that we don't know anything earlier than this point.

In fact, we can't even go as far back as the Big Bang, if we are being honest. If we take Hubble and Lemaître's expansion all the way back to the beginning then we have a singularity – and if there's one thing that theorists hate it's a singularity.

With vast mass and zero size, a singularity has infinite density and infinities are a pain. Their maths makes no sense, meaning every equation you chuck at them sends back gibberish.

Even something that's almost a singularity is too difficult to deal with at the moment. **Quantum mechanics** is a complicated, partially understood field that attempts to explain how subatomic particles interact in the Universe and it works very well the vast majority of the time, as long as you keep things small. **General relativity** is a complicated, partially understood field that attempts to explain how gravity works in the Universe, and it too works very well the vast majority of the time, as long as you keep things big. In a singularity you have tiny sizes subject to vast gravitational forces – and quantum mechanics and general relativity don't play well together. In order to understand the physics of a singularity, we would need to have a functioning theory of quantum gravity and, put simply, we don't.

So, the first moment after the Big Bang is definitely beyond us. How big a moment? About 10^{-43} seconds. It's called the **Planck epoch** and it is incredibly tiny: a period of time so small that, with our knowledge of physics today, it is too small for anything to have happened. Every physical interaction we are aware of takes longer than this to occur, so if time is just a measure of how things change, then such a short period may have no real physical meaning – that we know of.

Even then, according to current models we can't go back that far either. Some theories suggest that the first moment of the Big Bang was followed by the most violent and fastest expansion that the Universe has experienced to date, a time known as the **inflation epoch**. During this time, within the first 10^{-32} seconds after the supposed start of the Universe as we know it, the Universe expanded by a factor of 10^{26} in width. That would be equivalent to growing an atom to be a light year across in the tiniest fraction of a moment.

This idea of rapid inflation might solve a few issues with the Big Bang theory as a whole, such as why space seems so homogeneous and smooth or why the **cosmic microwave background** (we'll get to that) is similarly featureless. Taking such a small bit of the early Universe pushes any big changes in its structure outside our observable bit of it, meaning it's perhaps not surprising that space is similar in all directions. It's like zooming in on the surface of the Earth. Overall, it's a sphere, but go to a small enough scale and it looks pretty flat. There are those that have serious doubts about inflation, but for now it's generally accepted as true.

What triggered this temporary boost in the swelling of the cosmos may have been the breaking of the unification of two of the fundamental forces in the Universe.

We know of four fundamental forces: gravity, electromagnetism, and the weak and strong nuclear forces. Gravity affects things on vast scales and dominates the Universe as a whole. Electromagnetism produces everything from light through to electricity and is what binds electrons to atoms. The weak nuclear force is what makes radioactive elements radioactive and the strong force is what keeps protons and neutrons stuck together in the nucleus of an atom. They each have different strengths and ranges and in the world we see around us they are clearly quite separate.

One thing, however, that is sometimes difficult to get our heads around is that the world we see around us is very much a special case. It has a certain temperature and pressure – but none of these are true extremes and situations far hotter and far colder, far sparser and far denser, are all possible. At what we would consider to be extremely high temperatures, but perhaps the Universe outside of our experience might not, something weird happens.

At about 1 quadrillion degrees Celsius,[2] the electromagnetic force and the weak force unify. They act as one. The resulting force is known as the electroweak force and is a sign that in our Universe all the forces were once unified. Efforts to add in the strong nuclear force are known as Grand Unified Theories and are one step on the way to a Theory Of Everything, which would try to bring gravity into the mix as well. While no theories have so far been successful, finding them remains the holy grail of particle physics.

If true, as the Universe cooled in the early fractions of a second, first gravity and then the strong force would have split

---

[2] That's 1,000,000,000,000,000 for those of you who like long strings of zeros!

off from the others, becoming their own distinct forces. When the Universe cooled enough for the electromagnetic and weak forces to split, a vast amount of energy would have been released and may have been the trigger for our Universe's great inflationary leap forward.

We are now approaching one-trillionth (10^{-12}) of a second after the Big Bang and there is still very little in existence that we would recognise. Atoms did not exist – the Universe was just too hot. Had an atom been dropped into the early energy soup of the Universe, the electrons would have been stripped off the nucleus, the protons and neutrons would have separated and the individual particles would have been boiled down into their own constituent parts. Every proton and neutron, indeed many subatomic particles, are made up of even smaller bits called **quarks**. These come in six types, known as flavours – up, down, charm, strange, top and bottom – each with their own properties. If these names seem a little 'quark-number-four' to you, they were so called for a variety of peculiar and quaint reasons: some particles were perceived to be acting 'strangely', while others 'charmed' the scientists who discovered them.

Under normal circumstances, quarks can't be found on their own, only in pairs or more. The only known exception to this occurred now, from around one-trillionth of a second in. For the next fraction of a second, the entire Universe was filled with a quark–gluon plasma, where individual quarks floated around in a sea of subatomic particles, the other main constituent being another type of particle called a gluon, one associated with the strong nuclear force. These particles could only exist on their own because of the incredibly high temperature the Universe had at the time, up to a trillion degrees Celsius.

By about one-hundredth of a millisecond after the Big Bang, the Universe cooled enough for quarks to bind together into particles called hadrons, which thankfully included something more familiar: protons and neutrons. But it's at this point that something happened to change the course of the Universe again.

All particles of matter have a mass, an electric charge (even if that charge is zero), a spin and so on. Among the great zoo of particles out there, there are some particles that are almost identical, but have the exact opposite charge. These mirror-particles are called **antimatter**. The proton has an evil twin in the anti-proton, the electron in the positron and so on. While both particles act practically identically in some ways, they are the complete opposite in others, so much so that if you were ever to bring two opposing particles, such as one electron and one positron, close together, they would annihilate each other, releasing every constituent bit of energy and mass as a burst of light. As Einstein discovered in the early 1900s, energy and mass are in some ways interchangeable, but a small amount of matter equals a vast amount of energy, so the resulting detonation is vast given the size of the particles that are involved.

Understandably then, a stable Universe would need to be made almost entirely out of one or the other – either matter or antimatter – and not have large amounts of both. Luckily that's exactly what happened. The problem is we don't know why. Nominally, matter and antimatter should be identical in almost all regards. There should be only one real difference, their charge. When the Universe formed, it created both matter and antimatter in incredible quantities, which then promptly engaged in mutual destruction. But when the dust settled and the cosmic war was complete, matter was victorious.

Explanations for this are varied. The most likely and accepted theory is that there is some hidden asymmetry – basically there's something about antimatter, which we haven't yet discovered, that makes it very slightly less likely to be created. It doesn't even need to be a big difference – create 1,000,000,000 anti-particles and 1,000,000,001 particles in a small enough space and the fact that only one particle survives isn't a problem if it also happens in enough other places to result in the amount of matter we see today.

Perhaps a more interesting solution is that we just got lucky. Our bit of the Universe happened to have slightly more matter than antimatter but it could have been the other way around somewhere else. Because matter and antimatter can cooperate with themselves just fine, then it could be that, well beyond the borders of our observable Universe, there are entire solar systems and galaxies made of what we would refer to as antimatter. Perhaps they have their own antimatter aliens sitting in their antimatter chairs in their antimatter homes wondering why they ended up with only 'matter' in their part of the Universe.[3]

For whatever reason, by the end of the first second of existence, matter had become the dominant form of stuff in our neck of the woods and would go on to form almost everything the Universe would one day be filled with – well, the 'normal' stuff anyway.

The final act before the very first second was over was for one further particle, the neutrino, to release itself from the chaos. With the Universe so hot and dense, most things have been popping in and out of existence – spawning, colliding with something and disappearing again. But

---

[3] After all, referring to the material you yourself are made from as 'antimatter' seems like a significant sign of low self-esteem!

as things cool down and the Universe becomes less manic, some things are allowed to flow more freely. Neutrinos, a type of particle that you would need a barrier of lead a light year thick to (almost) ensure you've stopped it in its tracks, is one of the more elusive particles and so was one of the first to start to move freely within the nascent Universe.

One second into our Universe and already some of the most significant changes are complete. Our four fundamental forces have separated, the first building blocks of atoms have been formed and matter has won its war against its rival.

But the Universe has a long way to go. What we know of as the Universe today was still only a few light years across and as hot as around 10 billion degrees Celsius – far too warm for the first full atoms to form, let alone the first stars.

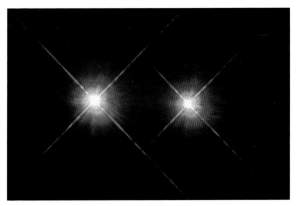

Alpha Centauri A, B and Proxima Centauri (too faint to be seen) are the closest stars to the Sun at a distance of 4 light years, though these are still incredibly distant even when compared to the most distant planets of our Solar System.
ESA/NASA

RS Puppis is one of several hundred Cepheid variable stars found in the Milky Way whose pulsating lights have been used to help map the extent of our galaxy.

NASA, ESA, and the Hubble Heritage Team (STScI/ AURA)-Hubble/Europe Collaboration. Acknowledgement: H. Bond (STScI and Penn State University)

While recording the brightnesses of stars in photographic plates, Henrietta Swan Leavitt recognised a pattern of variable stars known as Cepheid variables that would later be used to measure the size of the local Universe. She is pictured here (third from the left) at work with other women 'computers' at the Harvard College Observatory circa 1890.

Harvard College Observatory/Public Domain

Once centre stage in the argument to determine the scale of the Universe, the Andromeda Galaxy is now known to be an entirely separate structure to our own Milky Way some 2.5 million light years away, and not merely a nebula within it.
NASA/JPL-Caltech

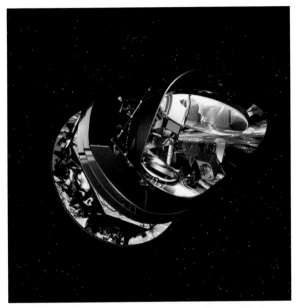

Peering through the Universe to see a form of light not too different from that used to heat a ready meal, the Planck mission has given us our best view of the cosmic microwave background yet.
ESA (Image by AOES Medialab)

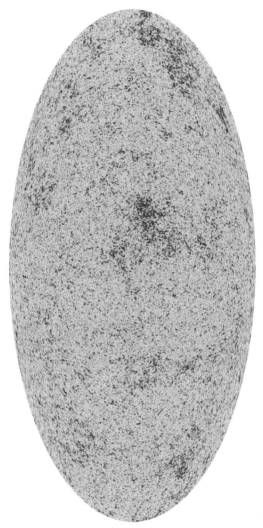

The hot and cold spots in this image of the cosmic microwave background, the light left over from the Big Bang, is the imprint of the foundations of the galaxies, clusters and voids that would later fill the Universe.
ESA and the Planck Collaboration

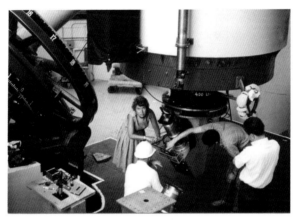

Along with Kent Ford, Vera Rubin provided some of the first evidence for the existence of the mysterious dark matter, a substance that outweighs ordinary matter five to one and yet is all but undetectable.
Carnegie Institution for Science

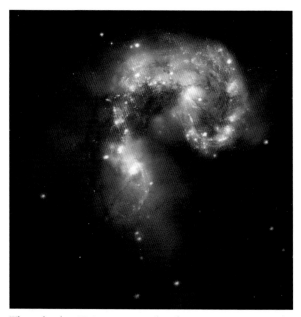

Though the Universe may be far more static in the modern day than it once was, it is not completely without change, as the Antenna Galaxy, the collision and merger of two separate galaxies, quite dramatically shows.

NASA/CXC/SAO/JPL-Caltech/STScI

The Cosmic Kitchen

I confess, I've never been much of one to cook for myself. With only a handful of exceptions when things have gone just right, I've usually seen cooking as a necessary evil, rather than a fun diversion. But even a relative cooking novice like me knows that to come up with a satisfying dish you need, at the very least, a suitable kitchen, the correct ingredients, a recipe to follow and time.

The Universe is no different.

The kitchen in this metaphor is the Universe. Overly hot and with no room to swing a cat, it's hardly ideal. But if it works in a small second-floor London flat, it'll work for the cosmos too.

When it comes to ingredients, the Universe is as bare as my cupboards before a weekly shop, but it won't take long to solve that.

Today atoms come in all sorts of sizes, from feather-light hydrogen to the heavyweight uranium and all the others in between. But at this point in the Universe, we weren't even close to that.

With the expansion of the Universe came rapid cooling and, about ten seconds in, the temperature was starting to become suitable for the formation of atomic nuclei (although technically the first atomic nuclei had existed for almost the entirety of the time that has passed at this point). The simplest element in the Universe is hydrogen – the full atom is a single proton coupled to a single electron. Take away that electron and you have a hydrogen ion. Granted it's no different to a solitary proton, mostly because it is a solitary proton, but technically it is the core of the most common element in the Universe.

But what about elements that are bigger than that?

Well by pure mass, the overwhelming majority of the elements we see in the Universe were formed by about the 20-minute mark in a process known as Big Bang **nucleosynthesis**.

Creating new elements either requires building up to them from smaller elements, a process known as **fusion**, or breaking down larger ones, called **fission**. Fission is easier because it requires much lower temperatures and can even happen spontaneously, hence why it is a major power source for us humans on Earth today, but there need to be large atoms around to break down. In the first few moments of the Universe, there were no such atoms.

In the modern Universe, fusion can only occur in the big pressure cookers that are the cores of stars, or in the brief explosions that end the lives of the largest of those stars, called **supernovae**, due to the intense temperatures and pressures required. But during its earliest moments, the entire Universe was hot and dense enough to act a bit like the core of a star and so fusion was happening everywhere at once.

Fusion couldn't have occurred sooner than this because the lightest product above simple hydrogen – called deuterium or heavy hydrogen – is the stepping stone to all major avenues for fusion but is also quite fragile. At temperatures higher than those experienced ten seconds after the Big Bang, the proton and neutron that need to stick together to form deuterium are easily broken up before enough time has passed for another reaction to occur and make the bond stable.

Even then, there were limits on what could be formed. Bigger elements require higher temperatures and the presence of a specific cocktail of nuclei to ensure that each step along the way is stable enough to survive the process before it is broken down again. As it turns out, you can't just chuck a whole load of subatomic particles in a mixer and end up with the entire menu of elements we see today.

By the 20-minute mark, it was done. The result was a mix of about 75 per cent simple hydrogen, 25 per cent helium-4 (the usual version of helium, with two protons and two neutrons to a nucleus), and some tiny traces of other things. That all-important deuterium was almost entirely used up, only 0.01 per cent of the leftovers were made of it, along with a similar quantity of helium-3 (like helium-4 but missing a neutron). Finally, a minuscule 0.00000001 per cent of the elements created was lithium, a very light metallic substance – and that was it! Though technically there will have been an atom or two of carbon, oxygen, nitrogen, iron and so on dotted around the place, they existed in completely insignificant quantities, even by modern standards where hydrogen and helium still dominate the show.

In fact, it is that dominance and the ratios of these primordial atoms that are among the reasons as to why we think the Big Bang theory is probably true. The relative quantities of all these elements are effectively dependent on one thing – the ratio of the number of baryons (protons and neutrons) to the number of photons (the bits of light providing the energy needed for fusion) at a particular stage in the Universe's existence. Because the Big Bang theory can provide that value, we can predict the quantities of each element that should be produced and compare our estimate to what we see in the Universe. The incredibly close agreement of our predictions and observations is a strong indicator that we aren't far off the mark in our thinking.

At this early time, with pretty much only hydrogen and helium in the Universe, the stage was not yet set for a complex, multi-course, gourmet meal of planets, moons and life – but as stars are made almost entirely out of these two elements, it was enough to at least make them. Not for a while though. The ingredients were prepared, but the recipe was still to be found. It would be tens of thousands of years before the Universe would change again.

During all this time, the temperatures were still too high for full atoms to form. Above a few thousand degrees Celsius, the electrons could easily be stripped from every atom in the Universe. What's more, had anyone existed at this time, they would not have been able to see beyond the end of their proverbial nose. With electrons flying around all over the place, no photon of light could fly any real distance before it crashed into one and was redirected. It would have been like trying to see through fog with bright headlights – there's plenty of light around but you can't see anything beyond a very short distance in front of you.

This would end with something called **recombination** – a slightly peculiar name given that things weren't combined before this. The cooling Universe eventually reached the point where electrons could attach themselves stably to atoms. In the course of a hundred thousand years or so, the Universe went from ionised and opaque to neutral and clear – the Universe was now transparent. We even have a surprisingly good grasp of when this process was complete: 380,000 years after the Big Bang. For the first time since the Universe formed, light could stream freely, and we were left with some of the best evidence of the Big Bang we have – the cosmic microwave background or CMB.

I've mentioned it before, that thing that was a little too smooth and featureless to be resolved without inflation. It's a signal, a glow of microwaves coming from all around us. If you happen to own an old television set that can be tuned to be between channels, then about 1 per cent of the flickering white and black static you can see is coming from these microwaves. Thankfully they're not quite the type of microwaves that vibrate water molecules in the device on your kitchen worktop, providing you with all the nutritional goodness of a well-made ready meal, but they're not too far off. This signal was found back in the 1960s by two radio astronomers who weren't even looking for it – serendipity and chance are surprisingly good scientists, it turns out.

Arno Penzias and Robert Wilson were working on a type of radio telescope in New Jersey in the USA when they noticed a faint source of noise. If it were audio, it would be little more than low-level white noise, annoying perhaps, but not exactly noteworthy. Rather than suffer a defective instrument, though, Penzias and Wilson tried all sorts of things to get rid of it but failed every time. At one point they thought it was interference from pigeon droppings in the receiver, so it's a good thing they decided to clean it out and check. The possibility of a Nobel Prize in Physics may seem like an odd way to motivate your kids to clean every now and then, but it's worth a shot.

Forced to admit this noise was real, they eventually realised they had discovered what theorists had predicted in the late 1940s: the leftover energy of the Big Bang.

When the Universe became transparent about 380,000 years after the Big Bang, the remaining light began to move freely through it. Just like a piece of metal heated up to high temperatures such that it glows, the colour and properties of the light came from the temperature of the Universe at that time. But that light has since been stretched by the expansion of the Universe in exactly the same way as had the light from distant galaxies that Edwin Hubble examined. What was once bright, visible light has been pulled through infrared and into microwaves over the course of the last 13.8 billion years.

Our view of the CMB has improved vastly as scientists have sent multiple satellites up into space to map it in its entirety – first the Soviet RELIKT-1, then a duo of NASA-led missions (COBE, the Cosmic Background Explorer, in 1989 and WMAP, the Wilkinson Microwave Anisotropy Probe, in 2001) and most recently the European Space Agency's Planck mission in 2009. You may well have seen images of the cosmic microwave background: a blue, yellow and red hodgepodge of stippled colours. But actually, that's not what it really looks like to us – I mean aside from the fact that microwaves are invisible.

The full microwave background has a number of other things superimposed on top of it. Our own galaxy and other nearby sources can be bright in microwaves, so to see just the 'cosmic' part, we have to first subtract an estimate of the contributions from all those sources. It's not the easiest thing to do.

Then we have the Doppler effect – the real one this time. Our planet, our Solar System, even our entire galaxy is moving through space at high speed. That means we are effectively travelling towards the source of the microwaves in one direction and away from it in the other. As these microwaves are a measure of the temperature of the Universe, it will make the CMB look slightly hotter in one direction and colder in the other, which also must be corrected for.

Even after this the CMB will look flat – very flat. The temperature of the overall CMB is about 2.725 Kelvin, or about -270.425 degrees Celsius. So, it's chilly. But the difference between one of those red-coloured spots and one of those blue-coloured spots is only about 0.0002 degrees Celsius.

Inflation helps explain why the differences are so small. Looking at opposite sides of the sky, we are looking at two parts of the Universe that have never had any form of contact. They are so far apart that light could not have crossed from one to the other, let alone have been in contact long enough for their temperatures to match up. Inflation removes the issue (sort of) by saying that the bit of the primordial Universe that our observable Universe grew from is so small that it isn't surprising the temperature is near identical everywhere.

Given the difference between a hot spot and a cold spot is about one part in 10,000 – incredibly tiny, in other words – they are clearly insignificant, right?

Well, no. Those tiny imperfections in the CMB were the blueprints for the structure of the entire Universe as we see it today. Without them, our Universe would be very different. Our recipe has been found.

Strictly speaking, the CMB isn't the recipe. It's the hastily scribbled-down copy of the recipe you make while watching a TV chef produce the dish faster than you can note the steps down.

The light from the CMB is a measure of which parts of the Universe are denser, or clumpier, than others. If the Universe had been perfectly smooth, then the CMB would have been perfectly smooth too.

But when inflation occurred, tiny, almost infinitesimal imperfections in the smoothness of the primordial Universe were blown up to incredible scales. Some bits had slightly more stuff in them, others slightly less. They would show up as hot and cold spots in the CMB. But they would also guide the Universe forward.

A denser spot has more gravity – it pulls on the material that surrounds it, making itself even denser and growing heavier over time, all while gravity makes it collapse in upon itself. The emptier spots with weaker gravity would be drained, fed upon by the clumpier bits of the Universe, like water being sucked into a sponge. The clumps get heavier and smaller while the emptier parts get emptier and larger. One day those clumps will form stars, galaxies, clusters of galaxies and so on. Meanwhile, the emptier parts will clear all but completely, creating the vast voids between galaxy clusters, interrupted only by the occasional galaxy that hasn't been sucked into the rest. It's the closest the Universe comes to a true vacuum.

In the tiny imperfections of the CMB is the imprint of what our Universe would one day look like, itself a reflection of the even tinier imperfections, magnified from quantum fluctuations of matter by inflation in the first tiny fraction of a second of our Universe's existence.

Not complicated at all.

The kitchen is now prepared for the last part of the preparation of our Universe. The CMB has shown us the recipe it was following, and nucleogenesis has given us the ingredients. If the Universe were patient, all that would be needed now is time.

But our observations of the Universe have shown that it didn't wait. For whatever reason it wanted a quick 30-minute meal, not an afternoon slaving away at the stove. To speed things up a bit, the Universe needed one more thing – and it wouldn't be easy for us to find.

The Dark Before the Dawn

That's not to say that what came next was quick by human standards. It would be hundreds of millions of years before the first stars were born (Figure 6). It just would have taken far longer if there hadn't been something else hiding in the dark.

And dark it was. Only a few million years after recombination, the light that was released had been stretched out of the visible range by the expansion of the Universe. It would be billions of years before it was the CMB we recognise today, but, even so, once pushed into the infrared the only sources of light in the Universe would have been beyond the ability of the human eye to see.

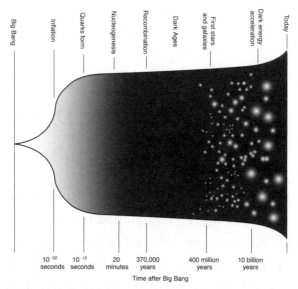

Figure 6: A brief history of our entire Universe to date, from the Big Bang, through a fast expansion known as inflation, on towards the formation of galaxies and stars that we can see in the present day. Much of the form of the Universe was baked in during the extremely early Universe, mere fractions of a second after it began to expand. The full impact of that would not be seen for thousands or millions of years, firstly when the light of the cosmic microwave background was released at recombination, and later when structures began to form.

One of the few sources of 'light' anywhere was the hydrogen that had been formed during nucleogenesis. Too cool to glow in visible light, it would instead have been a radio astronomer's dream, as it emitted almost solely in a type of light known as the hydrogen line, or the 21-cm line due to the wavelength of the light. This is still a type of light that dominates radio astronomy today. In a Universe far denser than the present one, the sky would have been ablaze with light our own eyes couldn't see, but a radio dish would have no trouble with. To have been able to witness the Universe slowly begin to separate out into clumps and voids, even if you had to see it in radio light, would have been fascinating.

The cosmic dark ages, the time between recombination and the cosmic dawn (the birth of the first stars) took a long time to pass – a few hundred million years at least. But, actually, that was surprisingly fast. Gravity may have been the dominant force in the Universe as a whole, but there was a lot going on out there that tried to slow it down.

One of those things is radiation pressure. Light, even invisible light, has a force behind it. When you stand facing the light on a sunny day, you are literally being pushed backwards by the radiance of the Sun. Of course, the force is tiny so it's not as if you need to brace yourself every time someone turns on a light switch. But on smaller particles, like the individual hydrogen and helium atoms that the primordial Universe was full of, the force can be substantial enough to change things.

Crushing gas creates heat. That heat makes light and that light pushes back against anything trying to crush the gas. Gravity will win – it always does – but the time it would have taken to do so on its own would have been vast.

Unless, of course, there was something out there that was heavy enough to attract and be attracted through gravity, but which didn't care about light.

Something called **dark matter**.

The evidence for dark matter doesn't originally come from the early Universe, but from the modern Universe and from a very unusual scientist: Fritz Zwicky. In the 1930s, he had appeared to show that clusters of galaxies were too heavy.

In the modern Universe, there is a mix of matter that is easily visible (stars beaming out light) and matter that's a bit trickier to see (cold dust and gas, planets, moons, etc.). Though galaxies vary somewhat, on average the ratio of luminous matter to dull matter will be about the same. So, if you measure how bright a galaxy is, you should have a rough guess of how heavy it is. You can also measure how heavy a galaxy is by seeing how it moves within a cluster of its friends. All the galaxies are moving under the influence of gravity, so calculate their speeds and apply some nifty maths and you have an independent measure of their mass.

Zwicky did both and found they didn't match. Not even close. The galaxies moved around inside a cluster as though they were 400 times heavier than the stuff that made them up! He claimed this must be due to matter that was not just dull but instead all but completely invisible – dark matter.

His idea didn't really get the attention it deserved at the time, though considering this claim came from a man who had once instructed a fellow astronomer to fire a gun out of an observatory to try and calm the air down that's perhaps not surprising.[4] Zwicky was undeniably brilliant, but also very inconsistent.

In the 1970s though, the proof would come. Vera Rubin and Kent Ford of the Carnegie Institution of Washington in the US were measuring how fast galaxies spin. If a galaxy is edge-on to you, one side will be moving towards you while the other side moves away from you. Due to the Doppler effect, this will make one

---

[4] It's actually slightly less crazy than it sounds.

side slightly blueshift and the other side slightly redshift.

They found that the galaxies were spinning too fast. The faster something spins, the more tightly it must hold on to itself to stop itself from flying apart. A Blu-ray disc will be perfectly fine if you put it into a functioning player, but attach it to a high-speed drill and you risk it shattering into very dangerous shards.

In a galaxy, gravity holds the stars in place. Spin too fast and gravity won't be able to maintain its grip. The galaxy will at the very least bloat as the stars move to wider and wider orbits, if it doesn't completely fall apart. And, according to Rubin and Ford, lots and lots of galaxies were spinning too fast for the material we could see to be held together. By their estimate, each galaxy would have to be about six times heavier than the stuff they could see.

This was rather less alarming than Zwicky's calculation of 400 times heavier, but that discrepancy can be explained almost entirely by the fact that Zwicky ran his experiments several decades earlier, when our best estimates for certain important numbers in cosmology were rather weaker than in the 1970s. When corrected to the modern values Rubin and Ford were using, Zwicky's measurements matched. He was right, which meant something else had to be wrong.

One possibility was that our understanding of gravity was wrong. Perhaps gravity acts differently over very large distances. It wouldn't be the only fundamental force to act strangely on different length scales. But some other observations are rather harder to solve that way – plus general relativity, our current favourite theory for how gravity works, has passed practically every test that has come its way in 100 years.

The other possibility is dark matter, some kind of stuff that we can't see, or have difficulty seeing, but is otherwise very heavy. There were two main competitors: MACHOs and WIMPs – because astronomy just wouldn't be the same without whimsical acronyms.

MACHOs are Massive Compact Halo Objects – basically normal stuff that's difficult to see. That includes brown dwarfs (failed stars), white dwarfs (dead stars), neutron stars (very dead stars), black holes (extremely dead stars) and so on. These could be clustered around galaxies in incredibly vast numbers but be so dim, or simply not be emitting light, that they become effectively invisible. Every test we've run, however, shows there isn't nearly enough normal stuff to account for so many dead or dying stars.

WIMPs on the other hand are the really interesting possibility. These Weakly Interacting Massive Particles seem to be an entirely new type of matter. They must have mass in order to provide the necessary gravity, but otherwise they are all but invisible. They don't appear to reflect or interfere with light, at least not very well, so we can't see them. They don't bash into other particles very often, so we also can't see their effects on other things, except through their gravity of course.

They act a little like a heavy neutrino – hard to catch, hard to see. Neutrinos can be found with vast detectors that wait for one of the trillion trillion particles that stream through the device to smash into something and produce a flash of blue light known as **Cherenkhov radiation**, alerting us to their presence. But similar experiments to catch dark matter particles in the act have so far failed.

This could of course mean that we are wrong about dark matter, be that its form or even its existence. Or it could mean that our experiments aren't big or sensitive enough to find it. For now, though, it's the best idea we have.

So how does it speed up the preparation of the Universe?

A particle that doesn't interact, or at least barely interacts, with light isn't going to care about radiation pressure. So, when gravity starts to pull everything towards those denser patches of the Universe, the matter may resist, but the dark matter slides on past and starts to bunch up. This makes the force of gravity even stronger, which pulls in more material, dark or otherwise, and so on and so on. The result is that the Universe rapidly becomes clumpy, much faster than if normal matter were doing this on its own.

The Universe was soon filled with a vast network of dark matter clumps connected by filaments – a **cosmic web**. Normal matter was pulled to the filaments and funnelled down them, feeding the formation of these new grand structures.

And by now all the pieces were in place for the finishing touches. The cosmic dawn had come.

A vast quantity of hydrogen gas all clumped in one place is the perfect starting point for a star. The earliest stars may have been huge, hundreds of times more massive than our own Sun and with the extremely short lifetimes that go with stars of that size. The nuclear fusion that powered those stars took hydrogen and stuck it together to create bigger and more complex elements – first helium, then carbon, nitrogen, oxygen and so it continued. With the incredible explosions at the ends of these stars' lives this material flooded into the rest of the Universe, enriching it with heavier elements that now make up nearly every object in our lives.

Early galaxies formed, mostly small and overflowing with the light from newly formed stars and the blinding explosions of their deaths.

At some point, exceptionally massive black holes came into being. It's unclear what came first – if the black holes preceded the galaxies or vice versa. Perhaps the early Universe was littered with black holes from the very beginning, parts of the Universe so incredibly dense that gravity crushed them into nothing before the Universe got going. Or perhaps, after their deaths, the earliest stars left behind black holes that are far heavier than those left by stars in the modern age.

What is clear is that the galaxies fed these black holes on dust and gas, causing them to grow slowly and mostly steadily for billions upon billions of years. Occasionally these galaxies would collide, merge and form bigger and bigger galaxies; the black holes in their cores would sink to the new galaxy's centre and themselves merge to provide one-time boosts to their growth. In the modern day, some of the largest galaxies out there are likely the products of many galactic mergers and every large galaxy now has at its core a **supermassive black hole**, up to billions of times the mass of our Sun.

Eventually, around 4 billion years after the Big Bang, a fairly nondescript galaxy would emerge, itself the product of many previous mergers. Inside, stars would be born, live and die, producing an increasing number of heavy elements like iron, gold, silver and uranium. A few billion years later, in the fading light of a supernova, made from the remains of at least the second generation of stars, an unimpressive cloud of gas would collapse in on itself. Among the range of stars that cloud created would be a small, dim yellow dwarf with a disc of dust and gas surrounding it. And from that disc, a tiny lump of rock and metal would emerge.

The Earth had been born.

A Note on the Present

It may have struck you by now that we are over three-quarters of the way through this book and we've only just reached something vaguely approaching the modern day. There's a good reason for that. Nothing much is happening any more.

The vast majority of what will change the Universe has already taken place. When the Universe was small and dense and incredibly hot, it was vibrant and active and changing all the time. Now, though, it is cold, dark and mostly empty.

If you read certain newspapers, you might be bombarded by reports of asteroids that scientists are 'worried will hit the Earth'. The fact is that nothing could be further from the truth. Asteroid impacts are possible and guaranteed if you wait long enough, but space is so big that the vast majority of the stuff that's out there will pass by harmlessly. Many of those headlines are for objects that are more distant than the Moon – about 400,000 kilometres away (or a quarter of a million miles). Is that a near miss by the Universe's standard? Absolutely. But practically speaking, it's way off.

In the same way, the incredible distances between stars, galaxies and clusters mean that it takes an incredibly long time for anything to happen.

That's not to say nothing at all is happening. Mergers of galaxies do still occur, though very rarely – the Antenna Galaxy, one of my personal favourites, is a relatively local galaxy that is doing just that. It's actually two galaxies that have made passes at each other before, like a bull charging down a matador, and they are now in the process of becoming one – like a bull and an unlucky matador.

On even longer timescales, the insides of the Universe are still slowly sloshing around. The complexity of this is perhaps best displayed in how many structures have formed and in their incredibly varied scales.

Around stars are solar systems: collections of planets, moons, asteroids, comets, dwarf planets and other detritus that orbit them. Those stars are often, though not always, parts of small clusters. Some are vast groupings known as globular clusters, where hundreds of thousands of stars formed all at about the same time. Others are so-called open clusters, much smaller groups of dozens or hundreds of stars. These latter types won't last long. Each star will slowly drift off into deep space leaving their stellar siblings behind, just like our Sun did 4 billion years ago.

These star clusters are bound together into greater complexes, galaxies, along with the gas and dust needed to make new stars. These galaxies vary from the very large, with trillions of stars in them, to the tiny dwarf galaxies that orbit their larger cousins. Some are unformed, like the blobby ellipticals; others have strong structure to them, like our own spiral Milky Way. And others still, like the Antenna Galaxy, haven't yet made up their minds...

Most of these galaxies are clustered together in groups. The Milky Way is one of the largest galaxies in the Local Group, probably just beaten by our nearest neighbouring large galaxy, the Andromeda Galaxy, but a fair bit bigger than the Triangulum Galaxy, the next in size. Dozens of dwarf galaxies round out our little corner of the cosmos. Within the cluster, each member orbits every other and, in around 5 billion years, our Milky Way and Andromeda Galaxies will go the way of the Antenna and have their own merger too.

This whole cluster is moving through space, itself orbiting within a cluster of clusters known as the Virgo Supercluster. This is an incredibly vast structure in the Universe, 110 million light years across and containing at least 100 galaxy clusters. And yet some studies show it may only be one part of an even greater supercluster, known as Laniakea. Centred on a point known as the Great Attractor, a huge gravitational anomaly that appears to be drawing much of the local Universe to itself, it is inconveniently located right behind part of our own Milky Way galaxy, so we know very little about it. There is even evidence that it may be part of yet another greater supercluster – though perhaps we'll leave that for now. Our full address within the Universe is complicated enough already.

Many of these structures can't be seen directly. The Great Wall, probably the largest known structure in the Universe at around 10 billion light years across, is known to us not because we can see the stream of galaxies and clusters that presumably make it up, but because **gamma-ray bursts,** a type of explosion that accompanies the deaths of the most massive stars, have been found at around the same distance from us in that region more frequently than almost anywhere else. We can see the intense light of these incredible explosions even when the galaxies they come from are all but out of reach.

There's so much more that could be said about the way we've studied our Universe and the incredible things we have discovered.

The expansion of the Universe has provided us with redshift, a method for measuring distances across the entirety of the observable Universe, and we've used it extensively to fill in the gaps in the map we made all the way back in the first chapters of this book.

The expansion of the Universe also messes with the way things appear as we look further back in time. We would normally expect a more distant object to seem smaller. What we find, however, is that beyond a certain distance, objects will actually appear to get bigger because when they emitted the light we are seeing, they were far closer to us and would have taken up a larger fraction of the sky.

Simple geometry, the accepted facts that squares have four right-angled corners and that the angles in triangles always add up to 180 degrees, may not be true over extremely large distances in space because space itself may be curved, another thing that takes a lot of imagination to get our primate brains to comprehend.

And then there's **gravitational lensing, baryon acoustic oscillations, quasars, string theory** and more.

Our Universe is a weird and complex thing, too complex to come remotely close to fully describing in a pocket guide.

But there is one last topic we need to cover. And it might not be fun.

The Bit You Probably Shouldn't Read

The Universe had a beginning. It will also have an end.

As strange and disconcerting as that may be, it's true. There is unfortunately no version of our current theories of the evolution of the Universe that allows things to continue for ever. However, in the same way that the Big Bang may not be the true beginning of our Universe, just the beginning of what we know today, the same may be true of the end.

Timescales get pretty extended here. Changes are dominated by the slowest and longest phenomena, simply because they are the slowest and longest. They are still going when the fast stuff has been and gone.

Stellar evolution can be an excruciatingly slow process. While the largest stars are born, live and die in just a few million years, stars like our own Sun take around 10 to 12 billion years to complete the same process. The smallest stars, red dwarfs, do the same thing in several trillion years. Not one red dwarf has died through 'natural causes'[5] in the history of the Universe and none will until around 100 lifetimes of our Sun have passed.

What's more, when stars die, the material that made them is released back into the Universe, able to make more stars. Our own Sun is thought to be at least a third-generation star. This may be somewhat reassuring. Even stars, the beacons of light in our dark universe, can be recycled.

---

[5] As opposed to, for example, demise as a result of an unfortunate encounter with a black hole.

Well, sorry to burst that bubble but it's not an infinite process. To produce light and heat, stars convert hydrogen into heavier elements. While that process is technically reversible, it would require gathering enough energy to split the heavier elements back into hydrogen again and, thanks to the most basic laws of thermodynamics (the study of heat and energy) that just can't happen.

Such a process would reverse something called **entropy**, a complicated concept that's most easily explained as a measure of how random or chaotic a system is. Imagine opening a bottle of blue dye in one corner of an Olympic swimming pool. We would expect the dye to spread out, slowly filling the pool and providing an unintentional Smurf cosplay for anyone unfortunate enough to go for a dip. This is order (the clear separation of dye and water) making way for disorder (the mixing of the dye and water, and an angry complaint from the *Avatar* impersonator). In scientific terms, the entropy of the system would have increased.

If instead a swimming pool full of blue water, naturally and without any external factors affecting it, happened to separate into clean water and a bottle filled only with blue dye, that would be order forming out of disorder, which is extraordinarily unlikely. Similarly, heat energy released into space can't somehow find itself being reabsorbed in such a way that would undo the process that created that heat in the first place – it's statistically unlikely and thermodynamically impossible.

The implication is that with every new generation of stars, more and more hydrogen will be lost. Eventually, the main fuel of the Universe will either be gone or be bound up in the most extremely long-lived stars, which, in turn, won't release much when they are done anyway. What's more, the frequency with which stars are forming across the Universe has been declining for a long time. It peaked around 10 billion years ago, a time known as 'cosmic noon', and there is no reason to believe it will ever start increasing again. In fact, evidence shows that, although star formation certainly isn't finished, more stars have been made in the past than will be made in the future. Like an endangered species whose habitat is slowly being removed, stars are heading towards extinction. Less than one-hundredth of the lifetime of a red dwarf has passed and already the sun is setting on the star formation of the Universe. Over the course of the next several trillion years or so, the Universe will slowly get darker and darker as the main sources of light die and fewer stars are born to take their place.

In the very distant future, there may be an even bigger problem. Most particles are unstable – this means that they break apart into smaller pieces if left to their own devices. But some are stable enough to hang around for a time, protons being the most important among them. If it turned out that protons, a constituent of every atom in the Universe, were secretly unstable, every atom in the Universe would eventually, and quite literally, fall apart. They wouldn't all fall apart at once, like a poorly made car in a slapstick comedy, though. Instead, they would fall apart randomly but in a statistically predictable way. Half would be gone within one **half-life** of the proton, half of the remainder would be gone in another half-life, half of that gone in another half-life and so on until eventually all useful matter in the Universe had disappeared. Experiments to try and determine how long this might take have so far failed, but we have been able to place lower limits on the half-life of a proton at 10^{34} years. In other words, it will take at least 10 billion trillion trillion years for half of the material in the Universe to self-destruct.

Of course, we could be lucky – perhaps protons don't decay at all. But that would not save us either.

There are two competing 'forces' that are important to the eventual fate of the Universe – 'forces' because neither of them are really forces at all, for different reasons. We've discussed both but have only given a name to one. The first is gravity. It constantly tries to pull things towards the centre of whatever mass is generating that pull. If the Universe were controlled only by gravity, then everything would be rushing inwards, collectively trying to clump together, crushed into a tiny space – sound familiar?

The other 'force' is the thing behind the modern-day expansion of the Universe: dark energy. Though it's tempting to associate it with the similarly mysteriously named dark matter, it is in fact completely different. There are really only two similarities between dark matter and dark energy: there's lots of both of them and we don't know what either is.

You can measure how much energy is contained in different components of the Universe. You can think of this as a gauge of what is the most important thing in the Universe at any one time. When the Universe was extremely young, within the first few tens of thousands of years, the amount of energy contained in radiation (light) dominated everything. This is unsurprising when you consider how hot the Universe was at the time. Eventually the Universe grew big, cold and dark enough that matter took over, with normal matter making up about one-sixth of that and dark matter making up the rest.

There existed throughout an initially tiny component, dark energy. It didn't get stronger or weaker with time, just stayed at the same level, waiting. About 4 billion years ago, with the Universe growing less and less dense, matter was becoming scarcer and dark energy became the dominant component. Today, about 70 per cent of the energy of the Universe is bound up in this mysterious dark energy, about 25 per cent in dark matter and 5 per cent in normal matter. In an incredibly dark Universe, light is barely even a factor.

Exactly what we think dark energy is remains difficult to describe. Space can somehow repel itself. It's as though empty space, a complete vacuum, is instead made up of vast numbers of tiny magnets all repelling each other. As the Universe grows and is forced apart there is more space where before there was less. This new space also repels itself, as though the magnets can replicate to fill the newly created room.

In fact, dark energy is currently doing the impossible – it is winning against gravity. Recent studies have shown that not only is the Universe still expanding in modern times, but that process is actually speeding up. The bigger the Universe gets, the more dark energy seems to dominate and the faster the Universe is forced apart.

Gravity is pulling inwards, dark energy is pushing out. The final fate of the Universe will be based on which 'force' wins.

If gravity succeeds, then the expansion of the Universe will one day slow, stop and reverse. The entire Universe will collapse in on itself in the direction of the singularity we may have had at the Big Bang – a situation known as the Big Crunch. For all we know, this could be a cycle. A Big Bang occurs, a Universe forms, expands, collapses and a new Big Bang occurs. Known as the Big Bounce, it's a possibility that our Universe is just one in a long chain of Universes stretching far into the 'past' and on into the 'future', with no Universe aware of any other.

If dark energy wins, then the resulting situation depends on how much it wins by. If it wins by a lot, then the Universe will continue to expand for ever at an ever faster rate. Paradoxically, this might actually mean that the observable Universe begins to shrink. Remember, the observable Universe is the bit that we can see. We've interpreted that as things beyond our horizon being too far away for their light to have had time to reach us. But, hypothetically, if an object is far too far away from us, then the light it emits can never reach us no matter how old the Universe gets because the space ahead of it is expanding faster than it can make up the distance. It's like the story of the tortoise and the hare, except the finishing line is on the back of a moving train. No one wins.

If the expansion of space accelerates hard enough, then one day it will be beyond our ability to see galaxies that are currently visible, as new light emitted by them cannot make the distance in time. Taken to its conclusion, the entire observable Universe would be the size of our galaxy, then our Solar System and ultimately just the Earth. The rest of the Universe would be invisible to us.

That's terrifying enough, but it wouldn't stop there. By now we would be mere seconds from the end. The observable Universe would continue to shrink, eventually becoming smaller than the distance between an atomic nucleus and the electrons around it. And it's not just light that can't cross the edge of the observable Universe – it's everything. No force would be or could be felt, so the electron would no longer experience the attraction of the nucleus. The atom would fall apart, followed extremely shortly by the nucleus itself, and then the protons and neutrons that made it up, as the observable Universe shrank to nothing, and space tore itself to shreds in the Big Rip.

Comfortingly, we very much doubt this will ever happen.

Far more likely, unfortunately, is that either dark energy wins by a little, or that the Universe is perfectly balanced and neither gravity nor dark energy can outmatch the other. In the first case, the Universe will expand for ever, but never fast enough to tear itself apart. In the latter, the expansion of the Universe will eventually slow and stop, and its size will be maintained for ever.

Either way, the fuel of every energy source will run out and the Universe will cool. The stars will go out and there will be no way to create new ones. Assuming protons don't decay, matter will either slowly radioactively decay to iron or fuse up to iron through the incredibly slow process of **cold fusion**. The remaining balls of pure iron will excruciatingly slowly collapse in on themselves, leaving black holes in their wake. Those black holes will slowly **evaporate** and, after no more than about $10^{10^{120}}$ years, the Universe will be completely dark, completely empty, and as cold as you can be – absolute zero. The Heat Death of the Universe will be complete.

Fun!

While evidence suggests the Big Rip probably won't happen, the other possibilities depend on a single measurement of the Universe, ominously and aptly called Ω, Omega. It is compared to a theoretical critical density, Ω_C. If Ω/Ω_C is larger than 1, that means the Universe is over-dense and it's time for the Big Crunch. If smaller than 1, the Universe is under-dense and the Universe will expand for ever with a Heat Death at the end. And if, and only if, the result is *exactly* 1, there will never be an overall winner between gravity and dark energy, and the Universe will one day stall – and there will still be a Heat Death at the end (Figure 7).

Current measurements of the density of the Universe place it almost exactly at 1, but the uncertainty on that measurement extends slightly above and slightly below.

What does that mean?

It means the Universe is a tease.

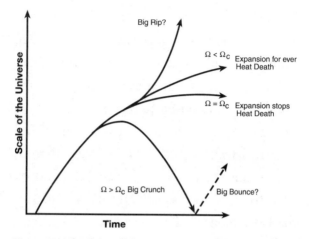

Figure 7: The fate of the Universe. Much of the future of the Universe depends on a single measurement: the density of the Universe, Ω, which is tied to the battle between gravity and dark energy. If Ω is greater than Ω_C, a calculated critical density, the Universe will eventually collapse into the Big Crunch, which may or may not then start the birth of a new Universe (the Big Bounce). If dark energy is far too powerful, it could result in the Universe tearing itself apart in the Big Rip. Finally, if Ω is either just less than or equal to Ω_C, then the result will be that all sources of energy will eventually run out in a scenario known as Heat Death. As it stands, none of these scenarios can be ruled out entirely, though the Big Rip is thought to be exceptionally unlikely.

How to End a Book That Just Ended the Universe

How *do* you end a book that just ended the Universe?

If you are looking for reassurance, you've probably come to the wrong place. Science deals with cold hard facts, and there's nothing colder and harder than the eventual fate of the Universe – well, nothing colder anyway... probably.

One point worth bearing in mind is that all but the most extreme theories place these 'ends' in the very distant future. Our Universe is 13.8 billion years old. The Solar System is about 4.5 billion years old. Animal life more complicated than a single cell is perhaps 600 million years old, depending on what you count as complicated. Modern humans are about 300,000 years old, recorded history is only about 10,000 years old and the *Lord of the Rings* films are about 20 years old.[6]

And yet, in comparison, the expected timescale for when the Universe will stop having stars suitable to sustain life is about 10 trillion years in the future, with the end of the Universe as a whole coming somewhere between a lot and a lot, lot later. The Universe has come a long way in the time it has had so far, and arguably it is already middle-aged. But its retirement will be very long indeed.

Perhaps it's better to focus on what we've already discovered and what is yet to be understood. In this book we've discussed particle physics, the fundamental

---

[6] That's most of the important milestones covered.

forces, cosmology, galaxy and star formation, evolution and much more, and we've only skimmed the surface. Human understanding far outstrips what we've discussed here and there's an incredible amount even our top scientists don't yet know.

For example, one of the most exciting events in the last few decades was the first detection of a gravitational wave. While the detection, the collision between two black holes in the medium distant Universe was not especially important for cosmology in and of itself, the detection of any wave was. It meant Albert Einstein's theories of relativity continue to hold strong and it has opened up an entirely new way to observe the Universe.

Unlike light, gravitational waves could potentially be visible from earlier than the time of the cosmic microwave background and may give us a way to see all the way back to inflation itself. It could be many years before our instruments are sensitive enough to detect such a signal, but the chance to prove right one of the most controversial parts of the Big Bang theory is tantalising.

Unfortunately, with topics as complex and fundamental as cosmology, there exists the possibility that there are some questions to which we will never have the answer.

What came before the Big Bang? What comes after the end of the Universe?

Does asking either of those questions even make sense?

Our understanding of physics to date tells us we won't ever know.

But since when has being told something is impossible ever stopped scientists from trying?

Glossary

antimatter – a set of particles that have identical properties to normal matter, aside from having reversed electric charges. When brought into contact with their 'normal' counterparts they annihilate one another and in turn produce a sizeable amount of energy.

baryon acoustic oscillations – ripples of matter left throughout the Universe by early motion, which can be observed today. These waves of material were carried by changes in pressure, similar to the way in which sound waves are made in the air.

blueshift – the apparent shift of colour seen in the light that is coming from an object approaching the observer. The compression of the light wave causes it to shift towards the blue end of the spectrum.

Cherenkhov radiation – a flash of blue light associated with the collisions and interactions of high-energy particles inside certain particle detectors. These detectors often contain a clear fluid, such as water, and light sensors around the dark space observe the energy and direction of the flash to determine the incoming particle's identity and direction.

cold fusion – the low-temperature version of nuclear fusion (see **fusion**). While atoms typically need to be at very high energies (temperatures) to fuse together into higher-mass elements, over extremely long timescales quantum tunnelling, a process that can cause particles to randomly jump past the barrier that typically prevents low-temperature fusion, can produce the same effect. Although cold fusion has also been seen by some as a potential method of energy generation on Earth, no path to making this process viable has yet been found.

cosmic microwave background (CMB) – the oldest light visible in the Universe, emitted when the Universe first became transparent around 380,000 years after the Big Bang. This light, which was originally visible, has been stretched by the expansion of the Universe into microwaves and is a valuable source of information about the conditions during the very earliest stages of the Universe.

cosmic web – the pattern of dense galaxy clusters, connected by filaments and surrounded by near-empty voids, that fills our Universe. Its structure was determined by tiny fluctuations of matter in the earliest moments after the Big Bang and it was sculpted in part by dark matter.

cosmological principle – the idea that no place in the Universe is 'special'. A combination of the observed fact that the Universe is homogeneous (the same throughout) and isotropic (the same in all directions).

cosmological redshift – the apparent shift of light towards the red end of the spectrum that comes from it being stretched during its long journey from distant galaxies to the observer by the expansion of the Universe.

dark matter – a theoretical material made of particles that interact poorly with light and matter yet exert enough of a gravitational pull to have affected the way the Universe has evolved. If our estimates are accurate, it outweighs normal matter five to one.

Doppler effect – the name given to the shift in frequency of a wave (sound, light, etc.) that occurs if the source and receiver are moving relative to one another (see **blueshift** and **redshift**).

eclipse – also known as a syzygy, an alignment of three bodies in space that causes one's line of sight to another to be blocked by the third. On Earth we sometimes experience solar eclipses, where the Moon blocks the Sun from the point of view of the Earth, and lunar eclipses, where the Earth blocks the Sun from the point of view of the Moon.

entropy – a measure of how unavailable thermal energy is within a system such as the Universe – simultaneously a measure of chaos or disorder in the Universe. As energy is radiated into the Universe, it becomes more spread out (disordered) and thus unusable. When all available energy has been dispersed in this way, the Universe will reach maximum entropy and all processes within it will effectively cease. There will be no stars or sources of heat left and so the Heat Death of the Universe will occur.

evaporate (black holes) – in this context, the process that causes black holes to slowly disappear over time due to something known as Hawking radiation. This theoretical process is extremely slow and would take 10^{64} years for even the smallest astronomical black holes to evaporate. The Universe has been around for only a little over 10^{10} years.

fission – the process where large atoms can, spontaneously or otherwise, decay into smaller atoms. Effectively the reverse of fusion. To date, this is the only form of nuclear power that has been successfully harnessed by humanity on large scales.

fusion – the process by which small atoms are combined into larger ones. The main power source of the stars. Can usually only occur at high temperatures or pressures (see, however, **cold fusion**).

gamma-ray bursts – exceptionally powerful explosions that occur at the end of the most massive stars' lives. They emit high-energy gamma-rays, along with lower-energy types of light (x-ray, visible, infrared, etc.) and can briefly outshine the gamma-ray emission of the entirety of the rest of the Universe.

general relativity – a field of physics that incorporates gravity and acceleration into **relativity**. The theory was first published by Albert Einstein in 1915.

gravitational lensing – the bending of light by a massive object (such as a star, galaxy or galaxy cluster), which causes the light to be magnified and distorted. One of the many implications of general relativity that has since been observed repeatedly in astronomy.

half-life – the time it takes for half the atoms of a radioactive source to decay. Equivalently, the time it takes for the rate of radioactive decay from a source (perhaps as measured by a Geiger counter) to halve.

heliocentric model – a model of the Solar System that places the Sun at the centre, rather than the Earth. Proposed or considered by a number of astronomers before him, but became a real possibility around 1510 with the work of Nicolaus Copernicus and published just before his death in 1543.

Hubble constant – H_0, a measure of the rate of expansion of the Universe. Its inverse gives an estimate of the age of the Universe.

Hubble–Lemaître law – the relation that describes the local expansion of the Universe in terms of the speed of recession of observed galaxies and their distances from the observer. At extremely great distances, this simple approximation stops working and more complicated models must be used instead.

inflation epoch – a brief (10^{-32} seconds) period at the very beginning of the Universe, during which it expanded faster than it had done before or since. This 'smoothed out' the early Universe and helps to explain why it is so homogeneous in the modern day. The idea is controversial but generally accepted.

nebulae – coming from the Latin for clouds, this describes a wide range of phenomena in the Universe, all of them in some way clouds of gas and dust. In early telescope astronomy, the optics were poor enough that many extended (i.e. not a 'point-like' or 'star-like') objects could look like a cloud, so star clusters and galaxies were included under this umbrella term until some time later. The debate around 'spiral nebulae', later called 'spiral galaxies', was central to the discussion of the scale of the Universe.

nova – a short-lived, bright burst of light that fades over weeks or months. Named after the Latin for 'new', the appearance of novae in the sky, seemingly from nowhere, made early astronomers think they might be the birth of new stars. However, novae are in fact from long-dead stars called white dwarfs feeding on material from another star until a critical point when the new material ignites. See also **supernovae**.

nucleosynthesis – the process of creating new atoms through fusion. In the modern Universe this occurs mostly in the cores of stars, or during certain high-temperature explosive events like

supernovae. In the first 20 minutes after the Big Bang, it also occurred throughout the Universe, producing most of the helium we see today.

parallax – the apparent shift in position of a nearby object against a distant background when observing it from two different perspectives. Also, the name of the distance-measuring method in astronomy that utilises this effect.

Planck epoch – the period of time (10^{-43} seconds) directly after the Big Bang during which our current understanding of physics is incapable of describing the conditions within the Universe. In many models of the early Universe, however, this lack of clarity may extend to after the inflation epoch.

quantum mechanics – the study of the interactions of particles and forces on extremely small scales. Most often relevant when discussing individual subatomic particles, but can, under certain circumstances, have important effects on the macroscopic scale as well.

quarks – the constituent building blocks of many subatomic particles called hadrons, of which protons and neutrons are examples. In the modern Universe they cannot be observed separately and are instead found in pairs or larger groups within their parent hadron. However, during the very early Universe, the temperature was high enough for them to float around freely in a quark–gluon plasma.

quasars – extremely luminous sources of light found in some distant galaxies. The light is emitted by a disc of superheated gas and dust found around a **supermassive black hole** in the core of the parent galaxy. Their light and extreme distance offers a way to probe the properties of the Universe between them and the observer and they provide a way to understand the process of galaxy formation and evolution, which is thought to be tied to the evolution of supermassive black holes.

recombination – a time, a few hundred thousand years after the Big Bang, during which the Universe cooled enough for electrons to attach themselves to atomic nuclei, forming the first stable atoms. This also made the Universe transparent for the first time, releasing the trapped light of the Big Bang, which eventually became the **cosmic microwave background**.

redshift – the apparent shift of colour seen in the light that is coming from an object moving away from the observer. The stretching of the light wave causes it to shift towards the red end of the spectrum.

relativity – the study of how things appear from differing perspectives. Classical relativity involves low speeds and has been fairly well understood since the time of Galileo (early 1600s). In the early 1900s, Albert Einstein included the concepts associated with the cosmic speed limit of light to extend this to high speeds in **special relativity**. Acceleration and the effects of gravity were included in **general relativity** a few years later.

singularity – in physics, a place of such intense gravity that our current understanding of general relativity and quantum mechanics is incapable of describing the conditions within. This is generally assumed to occur in the dense objects at the cores of black holes and, in some models, during the first moments of the Big Bang. It may also occur at the end of our Universe if the Universe's expansion ever reverses.

special relativity – a field of physics that incorporates extreme speeds (sizeable fractions of the speed of light) into **relativity**. First published by Albert Einstein in 1905.

string theory – an attempt at an understanding of quantum gravity, the interaction of general relativity and quantum mechanics, along with a number of other open questions in physics. It describes subatomic particles as vibrating, one-dimensional strings. Highly theoretical, extremely complicated and, amongst those who understand at least a bit of it, very controversial.

supermassive black hole – an extremely compact object with effectively zero size but a mass anywhere from 100 thousand to 100 billion times that of our Sun. Found in the cores of every large galaxy in the Universe and thought to play an important role in galaxy formation and evolution.

supernovae – either the bright explosion at the end of a massive (more than about eight times the mass of our Sun) star's life, or the bright explosion associated with the complete disintegration of a white dwarf. While once thought to be associated with classical novae (see **nova**), now known to be considerably more luminous, rarer and produced in very different ways.